How about we tell you the story
of the world... What do you mean you
don't have time? Wait, this...
will only take a few pages!
It's going to unfold before your eyes,
in this book, from the Big Bang
until today, from the
planet formation at
first men, from the
stone age, steam engines,
from the first cultivated fields to
the rise of cities and
civilizations, everything is connected and
the adventure leads to you!
We're going to embrace nearly 14
billion years of history.

So, are you ready?

Do you represent the universe in the making like a sphere of energy as big as an atom. All that will be created the reafter is contained in this tiny ball of energy. It's a...now, at this very moment that history begins.

For reasons we will never know maybe never, that ball explodes.

In one millionth of a millionth of a millionth of a millionth of a millionth of a second, it has gone from the size of an atom to the size of a galaxy!

The wave that is spreading and so is energy. The energy generated by this big bang and at the base of all the energy created in the universe. This energy will power everything from stars to living things.

When you refuel, you take the energy created during the Big Bang, the energy of the universe itself.
Now let's imagine that 380,000 years just

passed by.
You're going to be attending the
birth of our very first ancestors, the
atoms.

The first brick of the universe is the atom...
of hydrogen.
To create a universe, you need hydrogen, from
there, with the help of the heat and the
pressure, can create all kinds of pressure
of atoms.
The first atoms fuse into the primordial
universe, but they do not propagate not
uniformly, they accumulate in small pockets.

That's where the gravity factor, big sculptor of
the universe, will work his magic.

The formation of the first galaxies reveals an
immutable rule:
the meeting of matter and energy gives to more
complex elements.

Our modern cities are a good example of this.
Where there's something else can develop, if
there is nothing, there is nothing don't take
anything out.
The same is true on the scale of the universe.

300 million years after the Big Bang, at the
galaxies, clouds of dust are forming in the
galaxies contract under the effect of gravity,
which will result in an increase considerable
pressure and heat.

The temperature reaches 10 million degrees!
the hydrogen atoms then collide with each other.
violently and thus create a new element:
helium.
This fusion releases a colossal energy!

This is how the first stars are born.

All of a sudden lighthouses appeared in at night,
pouring their energy into the universe.
Let there be light!

There are now several billion stars but no
planets and no planet yet, no humanity.

Hydrogen and helium are not enough.
The complex and heavier elements that
allow the creation of solid objects
such as iron or carbon are manufactured in
heart of the stars.

Stars like our Sun are not only light sources,
they are also factories:
At its core, hydrogen becomes helium...
which in turn becomes lithium, and thus
that 25 of the most important elements
known to the universe including carbon,
oxygen, nitrogen and iron.

For example, more than 12 billion years ago, the
stars created the elements that made
possible the Iron Age, so the cities and the more
great monuments that mankind has built.

Let's take as an example the Statue of the Freedom.
Iron alone would not have been enough for the build. Its envelope is made of a element too heavy to be created in the heart of the stars: Copper.
Where does the copper in its envelope come from? Or the gold jewelry, uranium from nuclear reactors?

The stars don't have enough energy to manufacture them, obviously, the factory is not powerful enough! So, if we was blowing up?

Indeed, a few million years after the birth of the first stars, some of them exploded, they are called supernova.

Supernovas are the most explosive explosions powerful since the Big Bang, the energy the melting of more than one element heavy.

In the very furnace of their destruction, the stars create uranium, gold and all the elements that make up our world y including the copper.

The periodic classification of the elements and the library of the universe, everything that exists... comes from this sort of little chemist.
If it wasn't for the supernovas, we wouldn't be here,we're all stardust!

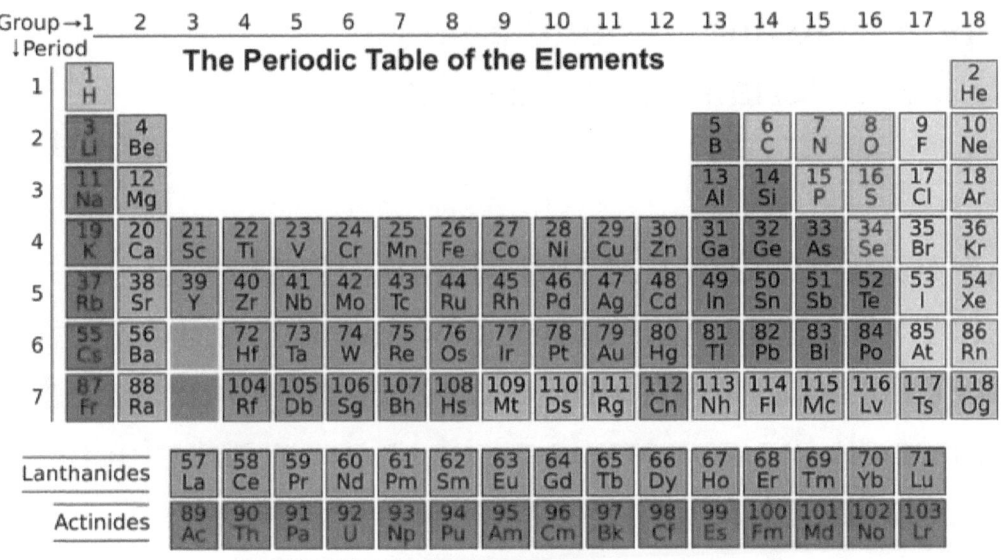

If you add the copper and tin, you get bronze,

without the supernovas, not from the Bronze Age.

When you buy a multi-vitamin juice, look at its composition you'll find copper, zinc, selenium, and all of them kinds of things that can't be made only by supernovas!

These elements are the seeds of future life and the ferments of its history, but our odyssey is just beginning.

For life to emerge, there must be an environment appropriate and to build a house,
all the materials necessary for the same place.

For the formation of the planets it's the same. thing, these are the materials that the universe has that will determine the types of planets and gather all the materials right in the right place takes a lot of time.

For 8 billion years the factories stellar will

continue their work creative.
Stars explode and are reborn and every generation produces heavier elements than the last one.

Then 4.6 billion years ago, a decisive event occurs: Birth of a super star, our sun.

It retains 99.9 per cent of the dust and of the gases of our solar system, but there are still still enough to make a few small things, planets. Ours will be the third from the star.

When the earth was formed 4.5 billion years ago of years, two thirds of the history of the Universe have already passed.
The first dawns are rising over a world hostile in the making.

The earth rotates so fast that the days only last 6 hours!

To imagine the earth at its origins, right after his

training, you have to imagine a any other planet. The sun illuminated a landscape of lava in fusion, in some places you could see masses of volcanic rock.

In this liquid rock the elements are all mixed up, but once again, the gravity will play a major role and put a strain on the order in this chaos.

The lighter materials date back to the surface, which will create a solid crust, and the heavier elements flow towards the centre of the Earth to form a metal core made of smelting composed of iron and nickel.

The movement of the liquid metal creates a magnetic field that extends into space and like a shield, it will protect our planet of emitted radioactive particles by the sun.

He's the one who will allow life to and later it will guide the explorers.
But first, Earth is going to do a bad meeting.
Four and a half billion years ago an object of

the size of March hit the earth at the speed of 40,000 kilometres an hour.

Our earth has absorbed most of the planetoid, but a cloud of debris at is bouncing around in space.
It will take a year for gravity to aggregate these debris.
They form a sphere that orbits around the Earth, a place she never left, the Moon was born.

The birth of the Moon is an event crucial in the history of the Earth, its creation has determined our current climate.

The moon stabilizes the earth and its orbit gravitational pull will prevent our planet to oscillate and will avoid the too abrupt climate change.

The impact also tilted the Earth's axis and it's because of this inclination that the seasons are possible.
The seasons are capital for the evolution of

the life and stability of the Earth's axis is therefore very important for the maintenance of this life.

The attraction of the moon also slows down the speed of the earth, which results in the longer days.

4,400 million years ago, was too hot for the water to exist at the liquid state but the atmosphere was saturated. of steam.

How do you get it to fall to earth?
Wants life to appear, it takes a little bit of rain! Temperatures are dropping.

For millions of years it's been raining... torrential downpours are descending on the planet.
It forms ponds, lakes, and then finally of the oceans.

3.8 billion, 800 million years ago, our planet has thus been enriched by a satellite, the moon and the oceans made their appearance, but she still

doesn't look like the environment we're familiar with.
For this, our planet must be
colonized and without a vital ingredient, no life forms could have appeared.

Three billion, 800 million years ago, under the the surface of the primitive oceans, a revolution has begun. 6 simple elements including oxygen, the hydrogen, carbon and nitrogen will be combine.

They will create substances that constitute all living things past and future.
The more spectacular of them and the DNA because in its spirals, it contains the secret code of life.

700 million years after the formation of Earth, life comes into play.
Our ancestors are not mighty giants but tiny organisms: Bacteria.

We are very self-centered, in fact we are

appeared very late!

The earth was the kingdom of the good bacteria before the animals and the men arrive!
You'd think the animals would have wiped out this old kingdom life, if that was the case.
We wouldn't be here!

Our body is a veritable zoo of bacteria...
Each of us contains more than bacteria than there are humans on Earth!

For billions of years, the microbes have had the earth to themselves.
Like the universe in formation, the first living things were small, simple and open at every possible opportunity.
But how did they generate all the existing life forms?

To understand this, let's go back to the beginning.

As we have seen, the energy contained in the

universe was created during the Big Bang and to survive every creature needs capturing some of that energy, and the more we can exploits it, the more complex you become.

Almost all of the energy that we consumption comes from the sun.

2.4 billion years ago a bacterium marine figured out how to harness the energy of the sun giving rise to the process of photosynthesis.

Photosynthesis produces a capital waste for the next part of our story: oxygen.
Soon, oxygen will redesign our world, but first of all, this bacteria has a another mission to fulfil.

The seas are then saturated with particles of iron and when iron and oxygen meet what's going on? It's rusting.

Rusted iron builds up on the sea floor for billions of years and will end up finally surfacing and, from this geological formation, man will derive the iron and steel.

It is these deposits that have allowed the industrial revolution.

Once there's no more rust iron in the sea, the bacteria will have to find a function to this oxygen,, they produce as long as the oceans are saturated!

The latter can only escape into the atmosphere. It is thanks to this that the earth has a completely different appearance from other planets.
Life will then take a crucial leap.

Some bacteria will learn to exploit the oxygen. Breathing is an old-fashioned way of
2.5 billion years, when life finds something that works, she's hanging on!
Oxygen is a game-changer. It's a source of extremely efficient energy!

This revolution paves the way for a whole

development series.

Over the 2 billion years that follow, the "living" becomes much more complex, the sky takes on a blue colour (from even as the oceans reflect it), the emerging continents, the earth takes on a face that we're familiar with.

550 million years ago, the earth celebrated its 4 billion years and the oxygen now make up 13% of the atmosphere.
Take a breath, you're going to speed up!

This is the Cambrian explosion, the version biological of the Big Bang.
The accumulation of oxygen paves the way to larger, more elaborate living things.

In 30 million years, a few seconds to universally, most groups of people in the of animals will appear.
The first fish with bones are colonizing the oceans. They don't look like us maybe not, but they're still our distant ancestors. If we have a

column spine, a mouth with a jaw and a spinal teeth, it's because of them.

We owe a lot to our ancestors fish. In fact, all vertebrates today are simple variations of these creatures!

For 4 billion years, plants and the animals remained confined in the marine environments, but an upheaval is taking place prepare.
Thanks to oxygen, an ozone layer is formed. form. It protects living beings from radiation.

Plants were the first to colonize the dry land, and then there's 400 million of years, the animals nested them on the not.

The first to discover the shoreline will be amphibians.
One of the most extraordinary moments in our evolution and probably the moment where the first amphibian came out of his primitive ocean, the moment he laid the first leg on the ground and took his first step on the breath of air.

Imagine our backsides grandparents get out of the water and fall in the in the middle of this fantasy world.
Look at all these trees all these bugs, there's plenty to eat! One day, man will colonize all the lands emerged.
But first, our ancestors must cut definitively bridges to the oceans.

Now, let's imagine that it's the season for love and just like frogs, frogs and amphibians are dependent on the environment aquatic.

That's where they lay their eggs.
But some will break free from their dependence on the marine environment, they will evolve and lay eggs capable of retaining moisture.
There's no need to go all the way to the ocean for find water, they carry it with them.

This is how amphibians become reptiles.

They could be 500 or 1500 km from the

coastline, and have enough water to the inside of the eggs to allow hatching.
The final link to their marine environment is cut, they can now colonize the earth.

300 million years ago, life continues to thrive in huge swamps tropical.

When plants die, the energy created by the big bang radiated by the sun is absorbed by it, is stored deep under form of coal. It's a gift that humans will exploit much later.

250 million years ago, the apocalypse came... and volcanic activity reaches a level never equalled since the creation of our planet.

The atmosphere is saturated with carbon, the extraordinary biodiversity of the Cambrian explosion is stopped dead in its tracks.
Near of all species are disappearing.

The extent of the Permian extinction is without

equal.

Extinction is a phenomenon recurring in the history of our planet, for 500 million years 5 cataclysms have eradicated the dominant species.

With each redistribution of new creatures have taken over.

This time, it's the dinosaurs' turn. These creatures ruled the earth for 160 million years.

This period marks the first appearance of conifers.

The moon ends up stabilize the Earth's rotation and days now last 24 hours.

At the beginning of the dinosaur era the continents are welded together to form a super continent called Pangaea.

But this territory is going to break up.

Africa is breaking away from South America, The Atlantic Ocean appears. It is a ,barrier that men will reach a day to go.

The stars of the day were undoubtedly the

triceratops and the tyrannosaurus but to their feet, we find tiny creatures and very important.

If we could extend our family tree so far, we've been would find very small mammals of the size of shrews surrounded by giant reptiles.

At that time, the mammals (which we make part), are a bit of an outsider.
They steal dinosaur eggs and are having trouble making ends meet, in fact, the dinosaurs prevent small mammals to evolve.

If there had been newspapers, they would have headlined "the mammals have lost the game."
We couldn't become much more big as a cat!

For 160 million years, every large, medium or huge animals were dinosaurs... That's a very long time! It takes us were in full swing! But the game is not not finished.

65 million years ago a broad object of 10 km,

probably an asteroid, hit the earth.

A cloud of dust blocked the rays of the sunshine, the temperature dropped and on earth creature weighing more than 25 kg has been found to be off.
It's the end of the reign of the dinosaurs.

By disappearing, the dinosaurs have made us a great gift and their extinction has been the signal of the mammals. Shortly after the end of the dinosaurs, the first primates appear.

They already have eyes placed on the front of the head which allows a better appreciation of distance. They also have skillful hands, like us, the fingers of the other animals are not configured like ours and don't have the opportunity to hold and handle objects.

Fifty million years ago, primates had evolved on a warming planet.
It is so hot that the poles are so hot that they are covered with vegetation. The continents

continue to drift.

In northern Africa, present-day Egypt is submerged by a primitive ocean.

The bottom of the sea and covered with small creatures called "nummulites". Their mineral shells made of calcium and carbon will accumulate on the funds sailors.
They're going to turn into layers of limestone.

It is this material which will be used much later to build the pyramids. Just observe the pyramids.

In detail the stones of some monuments to retrieve shells 50 years old millions of years!

10 million years ago, the earth had already a familiar face.

The Colorado River cuts through the Grand Canyon.
Mountain ranges, like the Himalayas, have risen.

They are already disrupting climate patterns.
The Isthmus of Panama emerges from the ocean.
It connects North and South America. South but
separating the Atlantic and the Pacific, it
modifies the circulation of ocean currents.

These geological events are preparing for the era
glacial. The planet is cooling.

Seven million years ago, our ancestors primates
lived in the tropics. But soon their territory will
be colonized.

This invader will have a major impact on
the history of mankind.
Despite its harmless appearance, it is indeed he
who will drive the primates to evolve. I will
explains; Grasslands are appearing
simultaneously over the entire surface of the
earth:
 The African savannah, the steppes of Eurasia,
the North American prairies, the pampas
Argentina, etc...
All these system echoes appear in at the same

time.

In East Africa, the savannah is gradually
replacing the savannah, forests, the traditional
habitat of the primates and trees are getting
scarce and our ancestors must then adapt.
They are from more and more in the same tree
and there's less food.

So they need to figure out how to get past
from one power source to another all
by changing "savannah zone".

So some primates go to the discovery of this
new habitat. This environment is best suited to
those who can walk on their hind legs.

With their heads sticking out of the tall grass,
they can be on the lookout for predators.
This posture is revolutionary because it free the
hands, and the hand will model the history of
mankind.

2,6 millions years ago, the first hominids

roamed territories sprinkled with silicon.
This element was created in the heart of the stars
several billion years earlier.

It's the second most abundant element of the
earth's crust and one of its characteristics and to
bind to oxygen for form crystals.
These combine to form a solid rock that can be
cut without being break.

The men started cutting stones more than 2
millions years ago and we've been continued to
use them for almost as many years!

By pruning just a stone, you'd get a ridge sharp,
and suddenly, the human being can do a
thousand times more things than before!
It's this little bit of technology that has made it
possible to for our ancestors to exist.

The flint and the key to the first revolution
human technology.
This is what we calls the Stone Age.
Hundreds of thousands of years later, another

amazing quality of silica will transform the world.

For example, our telephones, our televisions and many others. objects we use today.

But patience, we're not there yet.

The next crucial step in our history takes place according to a characteristic unknown to our planet, a power which she may be the only one in the universe.

Of all the planets and moons in the system solar, it seems that the earth is the only one or there is such a thing as fire!

Lightning and lava are good present on other planets, but to our knowledge only the earth possesses the resources needed for the fire to burn.

There is indeed a lot of fuel:

Plants, trees and an atmosphere rich in oxygen to fan the flames. If the fire had not been possible, we did not wouldn't be here.

Homo sapiens has taken a giant step forward!

800,000 years ago, our ancestors learned to fire

control.

This knowledge links us directly to the origin of the world.

All the energy available in the universe was created with the Big Bang and all life and a challenge to get some of that energy back.

The fire allows to cook food, the body then spends less energy crushing the food and can thus store more than calories.

Over time, the first men were and this additional energy is used to power the brains that grow accordingly.

The fire also paves the way for alltechnologies and soon we will transform ceramic clay, metal as a weapon and steam water.

Without fire, no combustion engine, no metal, it's a technology that opens up a new a vast field of possibilities!

200,000 years ago, modern man took form.

The larynx, which was located higher up in the throat of our ancestors, came down.
We can now articulate sounds more complex.
These are the beginnings of language and for the first time, individuals can exchange information and transmit it from a generation to generation.

You could say; my grandfather said that when there was no elephant, it was necessary to hunting zebras or my aunt told me that his cousin had found a watering place of across the river... Anyone can exploit this information and benefit from the experiences of the other.

Language allows us to move from stand-alone computer to that of computers networked.
 We share information!
We have no more need to experience things through yourself, from now on, you can borrow this experience to those around us.
Human intelligence then experiences a exponential development.
All the cards have been dealt, one new game can

begin.

100,000 years ago, man was a nomad.
With nimble hands and primitive tools, he can
communicate and he's a master of fire.
He will to be able to leave his African cradle
and take to new roads.

Continental drift has bound Africa to Europe and
Asia to create a super continent, Afro-Eurasia.

It occupies 85 million square kilometres.
Which is more than twice the surface area of
the Moon!
For the first men, this means that more than half
of the land are accessible on foot.
The dispersion of human beings to change the
give. We're one of the few apes that has
colonized more than one continent
simultaneously.

We are better protected against disasters that led
to the disappearance of large mammals and
that's a insurance against extinction.

But nature is testing us.
It's the beginning of the Ice Age.

50,000 years ago, glaciers came down of the
North Pole.
At the same time, the men continued their
expansion and reach China and Australia.

Thirty thousand years ago, Homo sapiens settled
in Europe and 20,000 years ago, at the height of
During the ice age, it reaches the cold tundra.
of northeastern Siberia.

Man has managed to overcome the rigours of the
extreme cold and develops the skills that make
him a creature apart.
The symbols that can be found by
example in the caves of Lascaux, are the
testimonials of new developments and of a
thought that goes beyond the mere moment
present.
It goes beyond our simple needs corporeal.
We can say with certainty that we're dealing
with a creature similar to us, that is with the

emergence of symbolic thought.

We are able to have before our eyes the
representation of a cow, and the whole the world
recognizes the image of a cow.

With the appearance of symbols, we know with
the certainty that we are in the presence of a
being human, creatures that see as
us and think like us.
From that moment on, the human race has
took a very different path from the others
cash.

At that time, an enormous amount of water
is trapped in the ice and the level of the seas are
falling.
The last barrier that has prevented human beings
from conquering the whole earth disappears.
Men go from Siberia to North America.

Man will now take centre stage, but he's good.
to recall that we represent a tiny part of the
history of the Universe.

To simplify matters, we will convert 14 billion years into 14 years.

In this hypothesis, the earth would only exist for the last 5 years, which is about the same as a third of the history of the Universe!
The great and complex creatures would have developed 7 months ago, the dinosaurs this would have shut down about three weeks ago, and the entire history of mankind would represent the last three minutes, the modern societies of the revolution would have started 6 years ago. Seconds!

In fact, the humans have only been here for one a brief moment on the scale of the universe.
It took many steps to reach this moment.

The ice bridges of the Ice Age have allowed the man to settle everywhere and the sea levels are rising.
The men are found isolated on two hemispheres. Each group is on its own condemned to make

the most of its resources.

As glaciers retreat, they carve lakes, rivers and bays. Map of the world takes on the face we know him to be.
In Africa, heavy rainfall makes overflow Lake Victoria and Albert and gives birth on the Nile.

In Eurasia, other rivers are being formed; the Tiger and the Euphrates in Mesopotamia, the current,Iraq, the Indus, in today's Pakistan, the Yellow River and the Yangtze River in China. And we're will see that these rivers will become very important after the ice melts.

It is on the fertile banks of these rivers that the first civilizations will be born.

At the end of the Ice Age the Earth was warms, plants and animals abound. Man can cease to be move to subsist, he becomes sedentary and the population is growing.
The big question is; how do we feed all those mouths?

Man must,find a way to increase the amount of food that he can get his environment.

A new discovery will change forever landscapes and the destiny of mankind.
We're going to learn how to plant seeds.

The plants we harvest are from the one that, millions of years ago, have allowed us to evolve.
Grasses are the unsung heroes of the world.
in the history of mankind.
A seed of grass is tiny, you can not feed on it and between hunting a buffalo and mowing grass, it's better to hunt for a buffalo!

And yet, the grasses are going to become the most important source of food.
But hunter-gatherers will be expecting thousands of years before cultivating them and do not will only use them when they have them. really need.

The best known species are the cane to sugar, wheat, rye, and barley. (All the cereals are grasses).

This family is more than just the beautiful lawn that we maintain on the weekends, it is also the staple food that all civilization depends! It is the majority of our calorie intake.
Once again, everything brings us back to the Big Bang.

Appropriating the energy that comes out of it, it's the eternal challenge of life.
Oxygen has changed the game. Fire has allowed to consume more calories and the transition to agriculture allows man to increase his productivity.

While hunter-gatherers had approximately 25 square kilometres of land are needed to provide food, cultivate a land will prove to be much more profitable.

The end of the cooler allows mankind exploit the sun's energy much more effectively efficiently.
It only needs a quarter of the time.

Of square kilometres to meet its needs and the most privileged region of everything is the fertile crescent in the Middle East.

Over there, we have a convergence of species at the plant and animal species susceptible to domestication.
As far as animals are concerned, there are cattle, pigs, sheep and cattle,goats.

On the vegetation side, there are varieties of wheat, barley, lentils, figs...
All concentrated in a very small perimeter.
On the other hand, Africa and the Americas do not have that very few wildlife species domesticated.

It's a crucial parameter! People who are fortunate enough to be in the right place have a great chance of becoming powerful and take a huge lead over the others.

Domestication will give rise to the exploitation

of the horse. This one gives a decisive advantage, from less to those who manage to train him.

Horses first appeared in America, but about 10,000 B.C., and then they disappear from the continent at the same time than other large mammals.
There were at least three species of horses in North America during the Ice Age.
They ranged in size from pony to percheron.

They lived there for 40 million years and then, after that gone! They're gone before the North American firsts and the opportunity to train.
Fortunately, before they disappeared, some herds had crossed the Bering Strait and conquered the steppes of Central Asia.

This fortunate coincidence, will have a formidable impact.
Around 4000 BC, the peoples nomads of Central Asia are learning to the domesticate.
We will soon exploit it all over Eurasia.

It is a powerful ally in the field as well as in the war.

The circle will be complete when 5,000 years more later, Christopher Columbus will bring horses during his second stay at the Americas.

These horses will be the first to pose the hoof on this continent since the disappearance of their species 10,000 years ago.

6,000 years ago, the transition to agriculture and the domestication of animals herald a new great mutation.

In space, gravity contracts the clouds of interstellar dust to form new stars and on earth, in Mesopotamia, new poles are coming into being also in place.

In Sumer, in the fertile crescent, the men establish themselves permanently. They can easily support themselves and new centres of power and of innovations are thus appearing.

Around 3000 B.C, the first innovations cities appear.
The Sumerian city of Uruk concentrates 50000 inhabitants on less than 2.5 square kilometres.
Thanks to the cultivation and breeding of the territory who could barely have fed a hunter-gatherer, can now support thousands of people.

But such a change is not without consequences and leads to a new addiction.
Once you've been through it, it looks like agriculture, we depend on 80 or 90 percent calories from one or two species!

In the Middle East, this is spelt and barley, as both cereals are harvested by the same season.
So we have all the food from the year in one go.
It's as if we got our salary once a year!
You have to put it in a safe place and plan your use.

If the harvest is bad, you end up with a famine

on our hands and we won't be able to get there. remedy before 12 months.

The management of these harvests conditions the organization of the first cities and for the counting man created the first system of writing. To protect them, he will train the first armies and to administer them, he invented the politics.

When hundreds of thousands of people live together, it is very difficult to ask for everyone's input, hence the need a government, a hierarchy political and social.

Men gather by the rivers like the Tigris and Euphrates, the Nile, the Indus, the Yellow River and the Yantze.
Civilizations are about to be born but first they must master the trade because the more goods they exchange knowledge with their neighbours, plus they develop.

Trade and long-distance communication

distance seems to be the necessary condition
the development of urban civilizations.

Surprisingly, the first civilizations will develop
thanks to an animal rather to discredit today: the
donkey.
Donkey caravans are both the highways and the
Internet of the time since road and foot prints
throw the foundations of the modern world.

They are not just vehicles for and goods such as
wood or bronze but also ideas and stories.
The civilizations they connect are the
first mentioned in the Bible.
They are converging on the Persian Gulf or the
ships take them to India.

The cultural and material exchanges between
these civilizations, are the beginning of the
globalization.
This trade is the key to understanding world of
yesterday and today because, like the first
civilizations, we exchange goods and create
networks and junctions are at the heart of these

networks. goods and create networks and junctions are at the heart of these networks.

When we occupy the centre of these operations, we are naturally at an advantage. The quantity of goods that transits through a region conditions its power and its importance.

Around 2000 B.C., the cabins are gradually giving way to immense constructions.
In Africa, the Great Pyramid rises up over the banks of the Nile in Great Britain Stonehenge comes out of the ground and in Sumer, the ziggurats, overhanging terraced buildings of a temple, rise up to the sky.

To seal the blocks of these gigantic structures, Sumerian builders use a substance that gushes out of the ground near the banks of the Euphrates. It's called bitumen.
Today, this fossil product is used for build the roads.

This is certainly the first product man-made oil.

While bitumen is highly prized, the viscous liquid that accompanies it annoys the Mesopotamians because it's too flammable. The elders calls it "Naft"; It's oil.

These vast fields of black gold will make cradle of civilization a land of wealth and, unfortunately, war.

But the Sumerians will leave us sometimes surprising inventions. Yes, they are, their digital system was not based on the number 10 but on the 12 and that's the reason for which the days are divided into 12 games, the hours in 60 minutes and the minutes in 60 seconds.

It's also likely that the Sumerians had invented the wheel, an invention that changed the destiny of mankind and who then allowed the tank to appear.
The Sumerian invention of the wheel allied to the
domestication of the horse by peoples nomads, is going to give this wonderful war machine!

Around 1200 BC, a vast conflict broke out.
It cuts off the copper and tin roads necessary to
manufacture the weapons in bronze.

Fortunately, the stars have designed an
alternative: Iron.
Blacksmiths make a crucial discovery.
At very high temperatures, they can to shape this
venerable metal.
Easier to sharpening and 700 times more
common than the copper, he's gonna rock the
world.
We are now entering the Iron Age.

In the first millennium B.C, the story's
accelerating.

After the Big Bang, the formation of the Earth,
the appearance of the first creatures and the birth
of man, the ice age has created bridges that have
allowed man to colonize every continent.

1000 years before our era man is scattered all

over the globe.

The trade road that link Eurasia and the Middle East.

North Africa have not yet penetrated the deserts nor crossed the oceans.

The peoples of sub-Saharan Africa and the Americas are isolated and have only few domesticated plants and animals, and by force of circumstance, remain imprisoned in a more primitive way of life.

600 B.C., the cavalry makes its appearance. For the first time, men are fighting on horseback because of course, once trained, these quadrupeds give an enormous advantage to their riders.

The alliance of the horse and iron weapons make the warrior formidable!

This technological breakthrough changes the the course of the battles, but it allows also the constitution of empires.

The succession of these empires are the new chapters of our history with in- 500 B.C. the Persian Empire, in -323 B.C. the empire of

Alexander the Great, -50 B.C. the empire Han Chinese, 117 A.D. the Roman Empire.

Empires change the balance of power and unified huge swathes of territory under the same control and with them of new religions are emerging.

This is the birth of Judaism as well as the Buddhism and Hinduism.
Despite the expansion of empires, some powers are isolated. Beyond the Himalayas, China remains cut off from the world.
but not for long.

In - 100 BC, an Chinese emperor sends a messenger to the west to alliances and the itinerary that it the road is the road that will later become the silk;
A huge commercial network linking the China, Central Asia and the Roman Empire.

The most surprising thing is that the chinese

merchants and the Romans did not are never met, at least not before.
The establishment of the Silk Road, but this business is about to explode.

Between - 100 and 200 years after Jesus Christ there will have three centuries of trade and cultural events of unprecedented intensity.
But the roads also carry threats invisible: Diseases.

Devastating epidemics may have contributed to the fall of the Roman Empire the Han dynasty in China.
At that time, new religious ideas
are spreading like wildfire.
In 312, the Roman emperor Constantine converted to Christianity and then three centuries
later is the birth of Islam.
At the height of its power, this religion will have unified a territory 2 and a half times larger than the Roman Empire! thanks to this hegemony, Arab trade will experience

an unprecedented development.

Arabs are at the crossroads of Asia and the Middle East.
Africa. Arab sailors return from China, Arab traders reach the Atlantic coast, finally, they are at the crossroads of all roads!

The secret of Arab trade is without no doubt the dromedary.
Like the horse, this animal's ancestors we flee North America via the bering strait. You should know that in one day a caravan of 6 dromedaries can be driven for a day,transport up to 2 tons of 100 km of freight!
In comparison, donkeys carry half as many in twice as long.

Dromedary caravans are opening up trade road in the vast desert of the Sahara and consequently the trade Arabic is spreading.
They're bringing salt from the Sahara to Rome, rice from East Asia to India, the Chinese secrets of paper manufacturing in Europe and allows for

the dissemination of many other inventions.

The Arabs brought a lot of food in Europe.
For example, oranges and lemons originate in
the South of China, and it's in the great Arab era
that they go reach the West.

In North Africa, a merchant's son the Italian,
who is also known as Leonardo Fibonacci, meet
Arab traders and take advantage of this to learn
a simple calculation method, but,ingenious, a
system born in India is used throughout the Arab
world.

Fibonacci's writings will disseminate this new
knowledge.
The unification of the calculation will facilitate
trade and commerce and to this day, we still use
almost all Arabic numerals.
Arab merchants will make known another
momentous invention born in China around 800
AD: Looking for an elixir for prolonging life, a
Chinese alchemist finds out how to give
instantly death by mixing carbon and sulphur

with saltpetre, a compound of the potassium, nitrogen and oxygen.

These elements, born in the heart of the stars, will now recombine to give the gun powder.

The recipe for gunpowder is a recipe for silk to win the Arab world.

The muslim soldiers use them to fight the Christians with cannons.

Later, the Europeans will in turn take hold of the invention and use it to perfecting their fire arms.

It is now 1492, the world has about 400 million inhabitants but it's still split in two.

In America, the Aztec civilizations, Maya and Inca are growing.

On the other side of the Atlantic, after the fall of the Roman Empire, Europe broke up into a single multitude of small independent states.

The Italian Christopher Columbus can now submit his project to different monarchs.

He finally managed to convince the great

sovereign of Spain to finance its shipment. The successful completion of the navigator's journey must much to the inventions that preceded it.

The triangular sails that explode effectively the strength of the wind are Arabic-inspired, the compass that guides it on the ocean is a Chinese invention and the magnetic field that orientates the needle has formed with the core of our planet.

Christopher Columbus is looking for a new road to reach India, and despite its stubbornness, he'll never find it, but he's gonna to reconnect the two halves of the world.

His discovery is a new turning point in history because before Christopher Columbus, these two major settlement areas were isolated from each other. When crossing the Atlantic, the navigator opens the rest of the world, that is, two huge continents, which are North America and of the South, with their millions of inhabitants and their resources.

Europeans and Asians are discovering a part of the world that was theirs completely unknown. One can easily imagine that before Christopher's Columbus trip, it was like they lived in twos different planets.

Trade roads born with the first civilizations of Asia, Africa, and the Middle East of Europe, are now crossing the Atlantic.
The centre of gravity of humanity is on the move and this network now create new crossroads and, once again, the powers that be are going to tilted.

For almost 2000 years, Europe has not played a major role, but with the discovery of Christopher Columbus, the West begins to take off and finally take the top because it is right in the middle of this new world out of proportion.

Food begins to circulate.

Corn of the Americas appears in Egypt and China and the Andean potatoes will be reveal perfectly adapted to the ground Irish and Russian.

The old cereal Fertile crescent like wheat begin to feed the Americas.

These new foods bring more of calories and energy.

Less than three centuries after the discovery of America's world population doubles to reach 900 million human beings!

But the inequalities between the two hemispheres will unfortunately lead to the worst atrocities.

The conquistadors (Spanish soldiers) want to take advantage of the wealth offered by the New World and arrive with firearms, horses and diseases and the result will prove to be an infectious disease and the slaughter.

In the years that followed, 95% of the

indigenous population of Latin America disappears as a victim of weapons and disease from Europe.
One thing's for sure, the the world will never be the same again!

Take for example the incredible story of the sugar:
Glucose is the only source of energy of our brain, which makes it vital. This commodity comes mostly from the sugar cane, and once again, this grass plays a crucial role in history of mankind.
This plant has been cultivated since 6000 in Asia.

During the Crusades, Europeans discovered sugar and take it home, but the problem is that it's very difficult to do
grow sugar cane in Europe. And there,on it, Christopher Columbus discovers America.

At first, of course, the Spaniards are very

interested in gold and silver, but what they're interested in want above all, is to get rich little no matter how!

Once the metals precious items have been looted, they start to plant sugar cane but do not want to under no circumstances work in the field and have therefore need for manpower.

Exactly, on the other side of the Atlantic they are can buy slaves and the main destination of the slaves of Africa are the sugar cane plantations. Sugar has undeniably contributed to shaping Middle Eastern history in the Crusades to the conquest of America not to mention the major role he played in the history of slavery!

Can you believe it?

At the beginning of the 18th century (300 years ago), man has made progress, but on a daily basis, he lives as simply as our farthest away ancestors.

The majority were farmers and practiced a subsistence farming. The objects of the were mostly made in the daily routine in small workshops.

It is necessary to then more than a year for the goods and the information goes around the planet, either longer than we have been it would take today to get to Mars!
This that impedes progress, it's just the working mode.

In Egypt, millennia before our time, 90% of the work was carried out by man and animal did the rest.

In 1700, in Europe, 70% of the work remains the work of man's hand, and yet, our physical strength is limited but a hidden ally in the depths of the Earth will to allow it to be multiplied: Coal.

This mineral is a fuel that replaces the wood efficiently, but its extraction will allow for a new breakthrough unexpected technology. To reach the coal seams, you have to dig deeper and deeper into the deeper.

The water invades the tunnels and it is therefore imperative to evacuate it.

In Britain, Thomas Newcomen in 1712 invented a pump powered by coal and propelled by steam.
It was the first steam engine. The combination of energy and motor, the fuel and the machine, will allow the man to go beyond the limits of his own physical force and change the world.
It's the beginning of the industrial revolution.

Associated with political revolutions that are shaking America and then France, it is this technological transformation that will change forever the course of our history.

Soon, the Atlantic zone becomes the new economic, and therefore cultural, pole, world's political and military and it is up to from that moment on, this one will dominate world geopolitics.

The companion ship now resonates from the

trains crash and around 1870 they invented
the internal combustion engine.
The Germans develop an application very
promising: The car.

And where does oil come back to this substance
which the ancients considered a nuisance,
becomes a commodity not to be missed.
It paves the way for further innovations in
the field of transport.

Telegraphs and telephones are the means by
which the information at the speed of lightning
and with the arrival of electricity in the 1900s
thanks to Thomas Edison and his current which
will be improved by Nicolas Tesla in 1920 with
its alternating current still used nowadays,
humanity has surpassed the night of his
kingdom.

In 1800, Europeans dominated 35 per cent of the
land and then a century later, they were
dominate 85% of them.
In the 20th century, with the advent of fossil

fuels and the the internal combustion engine, each event takes on an unprecedented scale y including war.
With technology, the military conflicts extend to a level international.

They take on an unprecedented importance.
In the 20th century, war will kill nearly 3,000 people tmes as many men as there were deaths in 2000 in history!

Industrial inventions and technologies invented in the 18 th century, reach their culminating in these events exceptional and that is the fossilization of micro-organisms in the swamps there are the millions and millions of years that he's had Permit!

The industrial revolution also brings with it the population explosion.
It took 200,000 years of history from the dawn of humanity at the beginning of the 20th century for that the population reaches 1,600 millions people and during the 20 th century, this figure

will almost quadruple!

The 20th century is a moment extraordinary in our history, but we have not never seen an upheaval of this magnitude before.

Today, we are now 7 billions and man has become an actor major on earth. We have learned to harness 50,000 times more energy than our ancestors.

10,000 years ago and that energy is fuelling the accelerated pace of our and the global network that criss-crosses the globe now the earth.

These virtual links are the direct legacy of the the first road laid out by our ancestors.

The seeds of the past have hatched into a present rich in energy and creativity and the web of
At the heart of this adventure is the immensity of the the universe, so vast that it remains a riddle.
At the heart of this mystery, each one of us is at

the both a living monument to the glory of the past and a story in the making.